iPhone 14
SENIORS GUIDE
2023

The Most Simple and Exhaustive Manual for the Non-Tech-Savvy to Master your New iPhone as a Beginner User

John Halbert

Turn on or turn off the hotspot 33
Make a new password 35
Change the name of the network 35

WI-FI 35
Enable Or Disable Wi-Fi 35
Sign up for a Wi-Fi network 36
Getting rid of a Wi-Fi network 36

BLUETOOTH 36
Turning Bluetooth on/off quickly 36
Connecting to a device 37
Disconnecting a device 37

5G CONNECTIVITY 37
How to switch from Wi-Fi to 4G/5G 37

Chapter 6: iCloud Account, Apps Payment, How To Setup Credit Card Or PayPal 39

Apple ID and iCloud 39
Sign in with Apple ID 39
How to use iCloud 41

Apple Pay 42
Setting up Apple Pay 43
Using Apple Pay 43

Chapter 7: Must-Have Social Apps And How To Set Them Up (Messages, Email, And Social Media) 46

Messaging on the iPhone 46

Social Media apps 51

Chapter 8: Detailed Instructions To Set Up The Camera And How To Capture The Best Photos/Videos 53

Capturing Photos 53

Recording Videos 57

Share Photos And Videos 58

Scan QR Code 58

Chapter 9: Health and Fitness 59

The iPhone 14's Health And Fitness Program 59

The Health App 62

The Activity App 63

Health Records And Fitness Data 64

How To Set Up Sleep Timers 67

Chapter 10: How To Use Dynamic Island (14 Pro And Pro Max Only) 68

Chapter 11: How To Setup Always-On Display 70

Chapter 12: How To Setup Widgets 72

Chapter 13: iPhone Accessories 75

    What Comes In The Box 75

    What You'll Need To Buy 76

Chapter 14: Apple Car Play, Music, Text To Speech, Reading, Etc. 78

    CarPlay 78

    Dictating Text-To-Speech On Your iPhone 80

Chapter 15: Best Tips And Tricks 82

Chapter 16: Maintenance And Battery Replacement Options 88

    How To Fix Battery Life Issues In iPhone 14 88

        Launch Your iPhone 88

        Turn Off 5G 89

        Use Low Power Mode 89

        iPhone Reset 89

    Battery Replacement Options 90

Chapter 17: FAQs 91

Conclusion 97

Introduction

In 2007, Apple introduced the new iPhone (officially released in June of that year). As with many groundbreaking innovations, the release of this device was surrounded by a great deal of hype, rumors, and speculation. Some heralded the iPhone as a great leap forward for technology because it transformed an iPod into a cell phone with internet capabilities. Others shied away from it because they saw it as nothing more than an overpriced status symbol. However, one thing was clear, the iPhone had changed the mobile technology market forever.

If you have only recently been introduced to the iPhone series, then this is the book for you. It's an easy-to-read guide that describes every step in great detail; you'll find it full of useful tips and tricks that will help you get the most out of your phone. Now you can benefit from all the special features that Apple has added to the iPhone 14 series like a veteran user.

The new iPhone 14 series is equipped with new hardware and software features. Among these is a new camera system that can stabilize action shots and the ability to shoot 4K at 30 frames per second (fps) in Cinematic Mode. Additionally, the new iPhones will have emergency SOS features to help users communicate in an emergency. This new technology will work in conjunction with SMS or iMessage and will be limited in length but will allow users to report an emergency or crime quickly.

Apple has unveiled two new safety features in the iPhone 14 series that should increase the security of users. First, the phone is equipped with a new satellite-based Emergency SOS, which uses the device's antennas to communicate with on-call emergency services. In case of a power outage, it will connect to a satellite so that help can be dispatched immediately. The new camera on the device also features a faster aperture and larger sensors to improve low-light photography and color. The iPhone 14 also features a new action mode that will help users make videos more stable.

The iPhone 14 comes with Apple's Cinematic Mode, which automatically focuses on a specific scene in a video. A fresh front camera has been added for your selfie needs. It features an autofocus sensor and a larger aperture. In addition, it has the strongest battery life yet, with up to 20 hours of video playback. This equates to about two days of impressive battery life.

The iPhone 14 Pro and the iPhone 14 Pro Max have slightly larger screens than the iPhone 13. The Pro model has a 6.12-inch screen, while the Max model boasts a 6.69-inch display. Both phones also feature a dual camera system for better video quality.

The iPhone 14 Plus features Face ID for unlocking and granting access to third-party apps. It also authenticates payments made through Apple Pay. Face ID has several uses across iOS systems and works with a series of sensors called Dot Projectors. These sensors project thousands of infrared dots onto the user's face to create a 3D scan of their face.

If you're new to the iPhone world or switching from an Android device, the iPhone 14 Seniors Guide is the perfect guide to get you started. It covers everything from basic usage and new features to differences between the iPhone models. It also gives tips on using the phone properly and getting the most out of it.

Chapter 1: Brief History Of The iPhone Since 2007

The iPhone has been available for several years but has not always been as colossal as it is today. Apple has released several versions of the iPhone over the years, but most have the same basic foundational features that make the iPhone unique. Apple has continued to add creative features to the iPhone over the years, adding 3D touch and distinguishing between a light and a hard press.

Apple's voice-activated assistant Siri became popular during the 2011 iPhone showcase, introducing consumers to artificial intelligence technology (AI). Other changes included the iCloud service, which allowed customers to store all their media on the Apple cloud, freeing up storage space on the phone. Another notable feature was the iMessage, which offered users a notification center, Twitter integration, and more.

Apple has reworked the design of the iPhone on several occasions, resulting in some significant changes in its look. The iPhone 1 featured a rectangular design with a 3.5-inch diagonal screen. This made it one of the first true handheld computers, and Apple claimed that the design was more durable than its spiritual predecessor, the iPod.

At first, Apple only made the iPhone available through AT&T in the U.S., but it was later distributed through various networks. In addition to AT&T, several other marketplaces sold the iPhone in Europe.

History Of The iPhone

The First iPhone – 2007 Model

In January 2007, Steve Jobs announced the first iPhone, combining a breakthrough internet communicator, a revolutionary mobile phone, and the iPod into a never-before-seen single handheld device. In addition, the iPhone had a 3.5-inch screen, a microphone, a multi-touch touchscreen display, headset controls, and some other unique features; some of these features laid the foundation of what came to be called the smartphone. However, the iPhone only offered internal storage of 16GB since it did not have the App Store. Moreover, the iPhone had a 2.0-megapixel camera and could only handle a 128 MB random accessory memory. Additionally, you could get the first iPhone 4GB model at a retail price of $499. As time passed, the iPhone sales spoke for themselves, proclaiming the original iPhone as the Best Invention of the Year.

The 2008 Model

The iPhone 3G, the first iPhone upgrade, appeared on the mobile market in July 2008, a year after the first model. The App Store was the most notable software addition. The new hardware allowed 3G connectivity and GPS tracking. The 3G feature enabled the iPhone user to access the internet easily without delays. In addition, the users could explore and download third-party programs through Apple's App Store, which as of the beginning of 2018, contained more than 2 million apps. Furthermore, like the first iPhone, the iPhone 3G had a TFT touchscreen of 3.5 inches, a 2MP camera, and 128 MB RAM. However, the WSJ stated that the iPhone 3G addressed the original's two primary flaws: its price and inability to connect to the quickest cellular networks. According to the article, the iPhone had the potential to develop into a trustworthy computing platform with broad adaptability. Nevertheless, with the addition of new hallmarks, the iPhone 3G 8GB model was made available at a retail price of $599.

The 2009 iPhone 3GS Model

In 2009, with the iPhone 3GS, Apple doubled the available storage capacity. The introduction of the App Store instantly changed everything. The 16 GB model was no longer enough for users to store all their photographs, music, and apps. Furthermore, Apple then upgraded the camera to 3 MP, adding the video recording feature. Although Siri wouldn't be available for a few more years, Apple also incorporated Voice Control. In addition, the iPhone 3GS had a 3MP camera, 256 MB RAM, and a 3.5-inch TFT touchscreen.

The 2010 iPhone 4 Model

The first iPhone to be made available on Verizon's wireless network introduced the world to several innovative features that the future iPhones would incorporate. The high-resolution Retina Display, the ability to multitask, the introduction of a front-facing camera, and FaceTime were all included. The iPhone 4 was a perfect choice due to its excellent performance, wide app selection capability, software and hardware fitness, and other details that improved its services in comparison to its predecessors. In addition, the iPhone 4 model had 512 MB RAM and could handle much more than iPhone 3G. Furthermore, it hosted maximum internal storage of 32 GB. Equally important is that the iPhone 4 introduced a new color variant, the white model, in less than a year. Lastly, it had the most extended lifespan among the other iPhones introduced.

The October 2011 Model

Although Steve Jobs (Apple's co-founder) oversaw the creation of the iPhone 5, which would be unveiled in September of the following year, he unexpectedly passed away a day after the announcement of iPhone 4S. Exclusive to the iPhone 4S at the time, would prove to be seminal for smartphones with the development of Siri, an AI-powered personal voice assistant. Furthermore, the new iPhone model also came with iOS 5, an overhaul of the previous iOS. In addition, Apple's first 8 MP camera with 1080p video capture made its debut with the iPhone 4S. To enhance the user experience, Apple also offered 64 GB of internal storage, but the RAM remained at 512 MB.

The 2012 - 2017 iPhone Models

The first Apple iPhone to go on sale in September was the iPhone 5, starting a trend that has persisted to this day. However, two years after the iPhone 5's release, Apple introduced two new additions to the lineup. In 2014, the iPhone 6 and iPhone 6 Plus were released simultaneously with the slogans "larger than bigger" and "the two and only." Additionally, the iPhone 6 Plus set the standard for successive display sizes, as the iPhone 7 Plus & 8 Plus models followed with 5.5" screens, as opposed to the 4.7" screens of the standard iPhone 6, iPhone 7, and iPhone 8 models. In addition to having more prominent displays, the iPhone 6 and iPhone 6 Plus had enhanced LTE and Wi-Fi connectivity, faster processors, and upgraded cameras. They then eliminated the 3.5mm headphone port and offered additional color variants, water resistance, and dust resistance in the iPhone 7 and 7 Plus.

The September 2017 Unveiling

The 4K Apple TV, Apple Watch Series 3, iPhone 8, iPhone 8 Plus, and iPhone X were all unveiled at the inaugural Apple event that happened at the Steve Jobs Theater located in Cupertino, California. A momentous occasion marked by the release of the iPhone X (in November that of year), celebrating the tenth anniversary of the introduction of the original iPhone. In addition, the iPhone X introduced edge-to-edge screens, wireless charging, and OLED displays for the first time in an Apple device. Furthermore, with the removal of the home button,

Face ID now enabled users to unlock their phones with just a glance, while the animated emojis unique to Apple products called Animoji mirrored their real-life movements.

The 2018 Models

Apple capitalized on the iPhone X from the previous year and introduced the iPhone XS, XS Max, and XR. The iPhone XS and XS Max included the most stunning displays ever, faster Face ID, a more intelligent and potent CPU, and a tantalizing dual-camera system. The third model, the iPhone XR, offered a brand-new liquid retina display that offers true-to-life color from edge to edge on the most enormous LCD ever used in an iPhone. Furthermore, the iPhone has permanently changed how we use smartphone technology, from FaceTime to Face ID. The well-received iPhone SE and C series were the offspring of these successes and developments.

The September 2020 iPhone Models

Four new 5G-compatible phones, the iPhone 12 and 12 Pro Max come in various colors. The new series has a 16-core Neural Engine, an A14 Bionic chip, and Ceramic Shield, which is highly efficient and has a four times higher drop performance than previous models. In addition, 5G's blazing speeds and low latency enables more responsive gaming, higher-quality video streaming, faster downloads, and real-time interaction. Furthermore, a Wide camera that now catches 27% more light and an Ultra-Wide camera that can now take photos in the dark are both brand-new to the iPhone 12.

The September 2021 Model

In September 2021, Apple released the iPhone 13 and iPhone 13 Pro, augmenting the company's sophisticated dual-camera setup. The iPhone 13, iPhone 13 Pro, iPhone 13 Pro Max, and iPhone 13 Mini have outstanding features, robust designs, the A15 bionic chip, and a significant improvement in battery life.

The September 2022 Model

The iPhone is considered the ambassador of smartphone users. In September 2022, Apple became the top tech company in the world thanks to its skillful creativity and advanced technologies. In retrospect, the future of mobile communication was not only pioneered by Steve Jobs but also remarkably influenced by Apple, Inc., and the present CEO Tim Cook. In September 2022, Apple released the iPhone 14, iPhone 14 Pro, and iPhone 14 Pro Max. The new models have a Super Dual Photon Sensor, a dual-pixel sensor with a higher camera aperture. To top it off, the camera can record 4K video at 30fps and take high-quality photos in low-lighting conditions.

Chapter 2: Different Models, Prices, And Specs

Pricing And Availability

iPhone 14 Pro starts at $999 with a storage capacity of 128GB, while prices for the iPhone 14 Pro Max range from $1,099 with the same 128GB storage capacity. More storage space may be rented out at an additional price.

Design

The iPhone 14 Pro models feature the same squared-off design as the iPhone 13 Pro models, which is reminiscent of the iPhone 4's design. The iPhone 14 Pro models have flat sides.

An all-glass front and a back that is textured matte glass are sandwiched in a surgical-grade stainless-steel frame when it comes to the design of the iPhone 14 Pro and 14 Pro Max. This design is essentially identical to the creation of the iPhone 13 Pro. The tint of the glass on the back coincides with the color of the stainless-steel frame, creating the illusion of streamlined simplicity.

Apple removed the notch of the iPhone 14 Pro models and replaced it with a new feature called the "Dynamic Island," which will be explained in more detail in the following section. The TrueDepth camera system is hidden behind the pill-shaped and circular perforations found on Dynamic Island. The pill-shaped cutouts and the circular cuts are combined into a single pill that is thirty percent smaller than the notch.

The top and sides of the phone each have antenna bands, while the bottom of the phone is where the Lightning Port for charging is located. Although there is no longer a SIM slot on the phone's left side in the United States, that particular location on the device still has the necessary circuitry for SIM cards in other countries. The front face of the handheld is protected with a durable display called Ceramic Shield. According to Apple, the Ceramic Shield is created by infusion of nano-ceramic crystals into the glass, improving the material's durability and enhancing its transparency. The Ceramic Shield is designed to provide an increased level of protection against scratches as well as ordinary wear and tear.

Since the previous year, the camera bump on the back of the iPhone 14 Pro models has expanded. This was done so that the iPhone could fit new camera technology. The camera has a setup with three lenses, each much larger than the others.

Sizes

The display size of the Apple iPhone 14 Pro is 6.1 inches. The Apple iPhone 14 Pro Max has a slight raise, measuring 6.7 inches. It is the largest iPhone that Apple sells, tied for first place with the iPhone 14 Plus. The iPhone 14 Pro and the iPhone 14 Pro Max feature dimensions that are considerably different from the iPhone 13 Pro and its versions. Although the iPhone 13 Pro is a little bit taller than the iPhone 14 Pro Max, the iPhone 14 Pro Max is a little bit more compact overall. Each generation is noticeably more cumbersome yet more robust.

The iPhone 14 Pro has the following dimensions: a height of 5.81 inches (147.5 millimeters), a width of 2.81 inches (71.5 millimeters), and a thickness of 0.31 inches (7.85mm). The height of the iPhone 14 Pro Max is 6.33 inches (160.7 millimeters), while its width and depth are 3.05 inches (77.6 millimeters) and 0.25 inches (6.4 millimeters), respectively (7.85mm).

Apple's iPhone 14 Pro weighs 7.27 ounces, equivalent to 206 grams, while the Pro Max weighs 8.47 ounces, equivalent to 240 grams.

Colors

Silver, Gold, Space Black, and Deep Purple are the hues offered for the iPhone 14 Pro and Pro Max. The hues Space Black and Deep Purple are new for the iPhone 13 Pro this year. Deep Purple takes the place of Sierra Blue as one of the color options.

Water Resistance

IP68 water resistance is included in iPhone 14 Pro and Pro Max. Like the iPhone 13 Pro, they can endure six-meter-deep liquids for a 30-minute timeframe.

The 6 in IP68 refers to dust resistance, meaning the iPhone 13 Pro can survive dirt, dust, and other particles. IP6x is the highest dust rating. The iPhone 14 Pro can endure accidental water contact, rain, and splashes. However, intended exposure should be avoided.

Apple says water and dust protection aren't permanent and may decline with usage. Apple's warranty does not cover liquid damage, so avoid submerging iPhone 14 Pro devices.

Display

The iPhone 14 Pro and Pro Max have flexible, upgraded OLED Super Retina XDR displays. A contrast ratio of 2,000,000:1 allows for deeper blacks and brighter whites, and a maximum brightness of 2,000 nits makes it easier to watch in full sunshine.

HDR enables 1600-nit peak brightness, which is brighter than the iPhone 13 Pro display. True Tone changes the display's white balance to the ambient lighting of the device for a paper-like viewing experience. Wide color support provides bright hues. There's also an oleophobic covering that resists fingerprints and Haptic Touch for haptic feedback.

Apple iPhone 14 vs. iPhone 14 Plus vs. iPhone 14 Pro vs. iPhone 14 Pro Max

	iPhone 14	iPhone 14 Plus	iPhone 14 Pro	iPhone 14 Pro Max
Display size, resolution	6.1-inch OLED; 2,532x1,170 pixels	6.7-inch OLED; 2,778x1,284 pixels	6.1-inch Super Retina XDR, OLED display, 2,556x1,179 pixels	6.7-inch Super Retina XDR, OLED display, 2,796x1,290 pixels
Pixel density	460 ppi	458 ppi	460 ppi	460 ppi
Dimensions (Inches)	5.78 x 2.82 x 0.31 in.	6.33 x 3.07 x 0.31 in.	5.81 x 2.81 x 0.31 in.	6.33 x 3.05 x 0.31 in.
Dimensions (Millimeters)	147 x 72 x 7.8mm	161 x 78 x 7.8mm	147.5 x 71.5 x 7.85mm	160.7 x 77.6 x 7.85mm

Weight (Ounces, Grams)	6.07 oz.; 172g	7.16 oz.; 203g	7.27 oz.; 206g	8.47 oz.; 240g
Operating System	iOS 16	iOS 16	iOS 16	iOS 16
Rear cameras	12MP (wide), 12MP (ultrawide)	12MP (wide), 12MP (ultrawide)	48MP (wide), 12MP (ultrawide), 12MP (telephoto)	48MP (wide), 12MP (ultrawide), 12MP (telephoto)
Front-facing camera	12 megapixels	12 megapixels	12 megapixels	12 megapixels
Video capture	HDR video recording with Dolby Vision up to 4K at 60 fps	HDR video recording with Dolby Vision up to 4K at 60 fps	HDR video recording with Dolby Vision up to 4K at 60 fps	HDR video recording with Dolby Vision up to 4K at 60 fps
Processor	Apple A15 Bionic	Apple A15 Bionic	Apple A16 Bionic	Apple A16 Bionic
Storage	128GB, 256GB, 512GB	128GB, 256GB, 512GB	128GB, 256GB, 512GB, 1TB	128GB, 256GB, 512GB, 1TB
RAM	Undisclosed	Undisclosed	Undisclosed	Undisclosed
Expandable storage	No	No	No	No
Battery	Undisclosed; Apple lists 20 hours of video playback; 3,279 mAh	Undisclosed; Apple lists 26 hours of video playback; 4,325 mAh	Undisclosed; Apple lists 29 hours of video playback; 3,200 mAh	Undisclosed; Apple lists 29 hours of video playback; 4,323 mAh
Fingerprint sensor	No (Face ID)	No (Face ID)	No (Face ID)	No (Face ID)
Connector	Lightning	Lightning	Lightning	Lightning

Headphone jack	No	No	No	No
Special features	5G enabled; MagSafe; water resistant (IP68); wireless charging; dual-SIM capabilities (e-SIM)	5G enabled; MagSafe; water resistant (IP68); wireless charging; dual-SIM capabilities (e-SIM)	Dynamic Island; Always-On display; 5G enabled; MagSafe; water resistant (IP68); wireless charging; dual-SIM capabilities (e-SIM)	Dynamic Island; Always-On display; 5G enabled; MagSafe; water resistant (IP68); wireless charging; dual-SIM capabilities (e-SIM)
Price off-contract (USD)	$799 (128GB), $899 (256GB), $1,099 (512GB)	$899 (128GB), $999 (256GB), $1,199 (512GB)	$999 (128GB), $1,099 (256GB), $1,299 (512GB), $1,499 (1TB)	$1,099 (128GB), $1,199 (256GB), $1,399 (512GB), $1,599 (1TB)
Price (GBP)	£849 (128GB)	£949 (128GB)	£1,099 (128GB)	£1,199 (128GB)
Price (AUD)	AU$1,399 (128GB)	AU$1,579 (128GB)	AU$1,749 (128GB)	AU$1,899 (128GB)

Table 1: iPhone 14 Comparisons (Source: CNET - https://www.cnet.com/tech/mobile/iphone-14-plus-pro-and-pro-max-comparing-price-size-battery-and-more-specs/)

Chapter 3: Basic Terminology About Main Functions

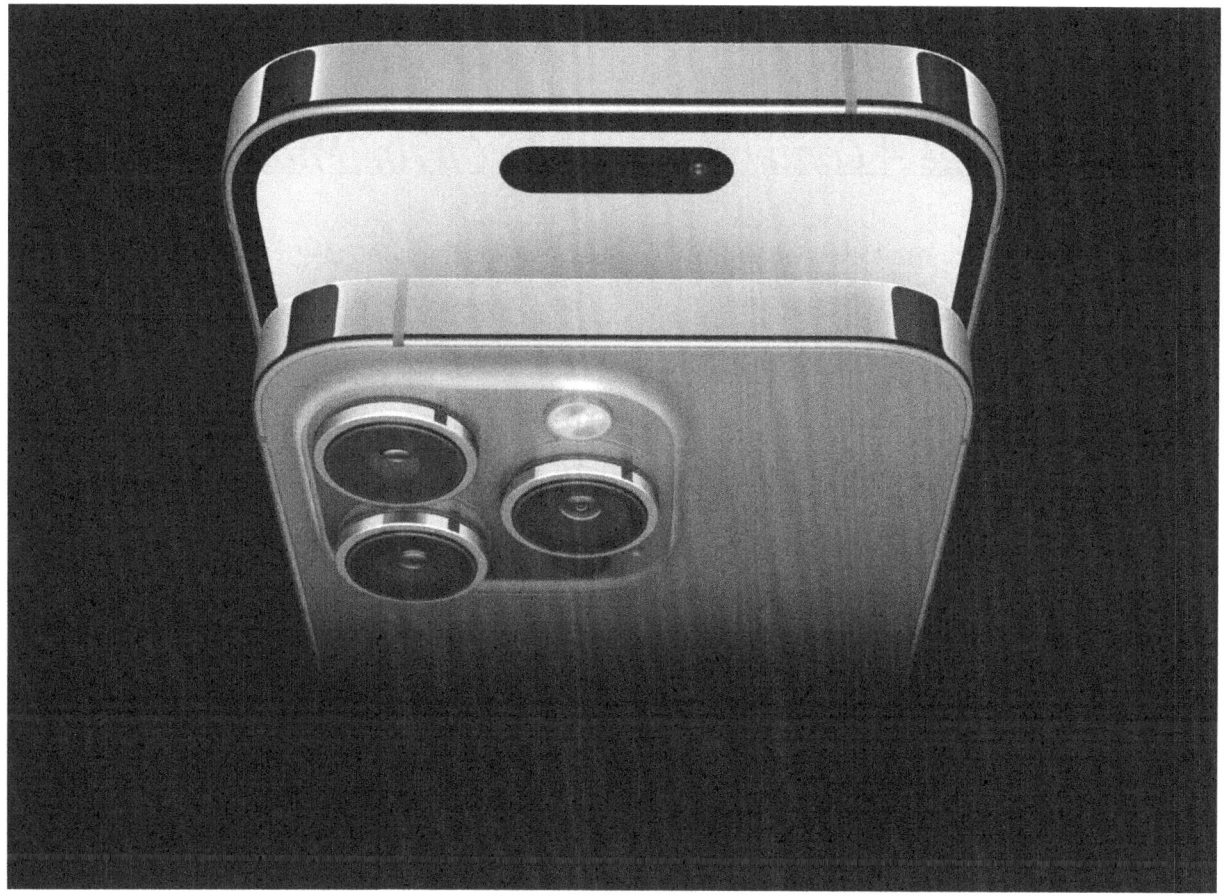

Ringtone, Wallpaper, Display, And Text Size

How To Change The Ringtone On Your iPhone 14

Let's begin by learning how to alter the ringtone on an iPhone 14. There are several built-in tunes available for selection.

1. Open the Settings app, then choose Sounds & Haptics.

2. Select the preferred ringtone to alter under Sounds and Vibration Patterns. For example, you may choose the call-ringing audio of your preference by selecting Ringtone.
3. You can test the ringtone by tapping or changing it, and then you can make it your new ringtone.

How To Change A Contact's Ringtone On An iPhone

You may select unique ringtones for your loved ones and special contacts by following the instructions below.

1. Enter the Contacts app. Find the person's name and tap it.
2. Select Edit in the upper right corner of the screen.
3. To choose a new tone, choose either the Text Tone or the Ringtone option.

How To Buy Ringtones Using iTunes

If the pre-installed ringtones on your phone don't fit your tastes, you may search the iTunes Store for an alternative and purchase it there.

1. Find the iTunes Store app.
2. Tap the More link in the lower right-hand corner.
3. Select Tones from the menu's top section.
4. Find the desired ringtone and then press the price.
5. Choose whether to assign the ringtone as a Ringtone, Text Tone, or to a contact. Alternatively, you can touch Done to adjust the parameter later.
6. To finish the transaction, you will need to key in your Apple ID password.

How To Create An iPhone Ringtone

You may create your own ringtone if you don't want to use the stock ringtones that come with your phone or spend money on them.

You may set any audio as a ringtone, whether a song or a recording of your child, as long as it is stored on your computer or in the Files app on your iPhone. Please note that tracks from Apple Music and other streaming services are protected and cannot be used. In other words, the audio file used to make the ringtone must be DRM-free.

1. Prepare iPhone audio file

Using the GarageBand program, you may create a ringtone for your iPhone. Please add the audio you want to convert into a ringtone to the Files app before proceeding.

The Files app provides access to iCloud Drive files. If you have audio files on your PC, you may upload them to iCloud Drive by visiting iCloud.com. This is how you can get to the audio file on your iPhone.

Additionally, you may be able to add music on your iPhone using an iPhone data transfer program like FoneTool. It will not delete any existing tracks on the device, unlike iTunes. The procedures to transfer music from a Windows 10 PC to an iPhone are outlined below.

2. Download FoneTool on your Windows PC.

Safely connect the iPhone to the PC by USB > When the program detects the device, choose Phone Transfer > From PC to iPhone.

3. Transfer to iPhone

- Click the arrow button and choose the desired music to transfer.
- Choose Songs.
- For music to be transferred to your iPhone, click Start Transfer.

Using GarageBand to Create an iPhone Ringtone

You must follow the instructions below to create your ringtone if the music or audio clip is already prepared for usage on the iPhone.

1. Install GarageBand from the App Store. Choose any instrument, like the Keyboard.
2. Tap the Project icon followed by the Loop button (close to the top-right corner of the screen).
3. Choose the required audio file (you may choose the audio file stored in the Files app by tapping Browse things inside the Files app.) > To import the audio file into GarageBand, press and hold it.
4. To cut the audio, tap on the track and drag both ends to the left and right. A ringtone should last less than 30 seconds.
5. Tap the Down Arrow button to save the song to My Music.
6. Find the clip under Recents and press and hold it > Tap the Share > Ringtone option.
7. Give the ringtone a name, and then select the Export button.

Select the "Use sound as" option if the export had completed successfully. The ringtone may then be configured as the default for a text message or a particular contact. To hear the new ringtone you created, hit "Done" and then open the Settings app.

Changing The Wallpaper

The wallpaper on the iPhone's Lock Screen and Home Screen can be changed. You may set your wallpaper in Settings or the Lock Screen's wallpaper gallery. Refer to "Customize your iPhone's Lock Screen."

Select Settings > Wallpaper > Add New Wallpaper.

Perform one of the subsequent:

1. Tap one of the icons at the top of the wallpaper gallery to add a photo, an emoji pattern, or other components to your background. Photos, People, Photo shuffles, emojis, and Weather are some available options.

2. Tap a wallpaper selection from one of the available categories, such as Featured, Suggested, or Photo Shuffle.
3. Select the "Add a picture" to your Lock Screen menu option to edit the image that acts as your background.

Tap Add, then follow one of the steps below:

- **Decide whether the wallpaper should appear on the Home Screen and Lock Screen simultaneously:** Select the option to set it as Background Pair.
- **Make even more adjustments to Home Screen:** Select Customize Home Screen. You can change the color of the wallpaper by tapping a color, tapping Photo On Rectangle button to use a custom image, or tapping the Blur button to make the background blurry and draw attention to the apps.

How To Modify The Display Size And Text

If you have problems seeing colors or other things, you may modify the display's settings to make reading the text on the screen easier.

Use display accommodations

Go to Settings > Accessibility > Display & Text Size to find text and display size options.

Select one of the following options to modify the text:

- **Bold**: Shows the specified text with boldfaced letters.
- **Larger Font**: Enable Larger Accessibility Sizes, then use the Font Size slider to modify the text size.
  The option lets you change the size of the fonts in apps supporting Dynamic Type, like Calendar, Messages, Contacts, Settings, Mail, and Notes.

- **Reduce Transparency**: This option decreases the background's transparency and blurs it.
- **Increase Contrast:** This option enhances contrast and readability by modifying color and text formatting.

To increase or decrease text size, use the slider

- Tap the All Apps icon on the screen's bottom to alter the text size for every app on your smartphone.

    When not in use, the iPhone shuts off the display to ensure the battery power is conserved, locks due to security reasons, and automatically goes to sleep. The iPhone may be instantly awoken and unlocked when it is time to use it again.

The iPhone And The Internet

Connecting The iPhone To Wi-Fi Networks

Navigate to Settings > Wi-Fi. Here you can enable Wi-Fi and connect to available networks.

Select one of the following from the list below:

- A network: If prompted, provide the password.
- For you to join a secret network, you need to type in the network's name, security type, and the password.

Joining A Personal Hotspot

You may use a cellular internet connection from another iPhone or iPad.

1. Select the device's name sharing the Personal Hotspot by navigating to Settings > Wi-Fi.
2. If prompted, enter the password you find in Settings > Cellular > Personal Hotspot on the device that is sharing the Personal Hotspot.

Connecting Your iPhone To A Mobile Network

If a Wi-Fi network is unavailable, your iPhone automatically connects to the cellular data network provided by your carrier. If the iPhone is not connecting, verify the following:

- Go to Cellular under Settings.
- Confirm that Cellular Data is enabled. Tap Cellular Data. If you are utilizing the Dual SIM facility, you must be aware of the line that you have selected. (Only one line may be used for cellular data.)

The iPhone conducts the below actions until the connection is made when an internet connection is necessary.

- Attempts to establish a connection using the most recently used accessible Wi-Fi network.
- Presents a list of accessible Wi-Fi networks, from which the user may pick one to join and then connect to that network.
- Establishes a connection with the mobile data network provided by your carrier. 5G mobile data may be used in place of Wi-Fi. If so, the phrase "Using 5G Cellular for the Internet" will appear next to the name of the Wi-Fi network you are now connected to. Press the Info button next to the network name, and tap the Use Wi-Fi for the Internet option. This will revert you back to Wi-Fi.

Chapter 4: How To Setup The iPhone 14 For The First Time

Your iPhone 14 has finally arrived. The next step is to set up your phone, which might be hard to do without help.

Before Setting Up Your Latest iPhone

If this is the first time:

Make a copy of the old iPhone you have: If you are switching from a previous model of the iPhone, you can move data from the old phone to the new iPhone 14. You need a backup that's been kept up to date.

Hold on to your previous phone: You should have your old phone ready. This will assist in speeding up the set-up process.

Also, bring your charger: Even though the new iPhone 14 may have an adequately charged battery to go through the setup procedure, it is helpful to keep a charger ready.

Check for Internet access: To configure your iPhone, you must first access a Wi-Fi network.

Prepare Apple ID information: Keep your email ID and password credentials ready.

The Set-Up Process

1. Turning on the iPhone.

Hold down the right-side button. Within seconds the Apple logo will appear. After some time, the Hello screen appears. To get started, swipe up.

2. Follow the on-screen instructions.

You may be required to perform some minor prompts on the iPhone. Fortunately, everything is simple. To begin, you must:

- Select a language. English should be at the top if you live in the United States.
- Select your home region or country. Again, "United States" should be at the top of the list if you are in the United States.

3. Install Quick Start on your iPhone.

The Quick Start screen will appear on the following page, allowing for automatic set up of the new iPhone 14 using an older iPhone. If you choose this option, you must turn on the old device, use Wi-Fi to connect, and bring it close to your new iPhone 14.

- There will be a prompt for authentication or connecting your two phones. It is possible to use a verification code or scan a pattern from the old phone.
- If you decide to perform the process by hand, proceed with the following steps.
- Connect the new iPhone to your home network's Wi-Fi.
- Set up your eSIM. You may transfer from an older phone (which prompts you to press the side button twice) or do the setup later in settings.
- Face ID and Touch ID can both be enabled.
- You have complete control over how your data is transferred. You can do it using iCloud or your old iPhone.
- Let your devices remain powered as this may take longer depending on the amount of data or apps on the previous phone. Complete the procedure and the iPhone 14 boots up on the iOS 16.

4. Manually configure your Apple iPhone 14.

Select Set Up Manually that appears on the Quick Start display:

- Have a connection to a wireless network. Allow a few moments before using your phone.
- View the data and privacy prompt. When you're finished, click the Continue button.
- Install Touch ID or Face ID. They will enable you to unlock your phone using your fingerprint or face.
- Come up with a six-digit code for unlocking the iPhone 14. It serves as an alternative method for getting your iPhone to work. Enter your code twice to confirm.
- Choose the method of recovering data and apps. You have the option of restoring from a Mac/PC, iCloud backup, transferring data directly from an iPhone or Android device, or not transferring data and apps at all.

- Depending on how you restore data and apps, you may need to log in to an Apple account using your iCloud credentials. A verification code may be required to prove it is you.
- Next, accept the terms and conditions, make changes to your settings, update to the most recent version of iOS, and, if necessary, configure Siri, Apple Pay, and Screen Time.
- Set up True Tone display, and select a dark or light mode and your view, whether zoomed or standard.

5. The iPhone 14 is now ready to be used.

Swipe up when you're finished. Give your device time to transfer the data and apps.

Wake Your iPhone

- Press the side button
- You can disable wake-up in Settings > Display & Brightness
- Tap the display feature

Unlock the iPhone using Face ID

- You can tap or lift the screen on iPhones to wake them up. Simply look at the iPhone. Make sure your face is visible to the front camera.
- Lock icon changes from closed to open once you unlock.
- Swipe up starting from the screen's bottom.
- Press the side button to lock the iPhone.

Chapter 5: Internet Setup, Use, And How To Switch From Wi-Fi To 4G/5G

Personal Hotspot

Set up your device as a hotspot to share your data connection with other Wi-Fi-enabled devices.

Turn on or turn off the hotspot

- Go to Applications > Settings to get to the settings app.

- Use the drop-down menu to choose "Personal Hotspot." Next, choose the personal hotspot on/off switch from the menu bar to turn on a mobile hotspot.
- It's important to remember that you have to turn off Wi-Fi to use the Personal Hotspot.
- Move up from the screen's bottom to access the Control Center, which enables you to activate or deactivate your mobile hotspot easily. To do so, press and hold the middle of the Connectivity tab for a few seconds after making the swipe. Examining the icon of a personal hotspot will lead one to the location of the hotspot.

Make a new password

To change the hotspot password, go to the Personal Hotspot menu and click Wi-Fi Password. Then, alter the password as you like. Click Finish when you're done.

Change the name of the network

Scroll to the bottom of the page and click General > Info > Name while in the Settings panel. Change your device's name as may be needed.

Note that the name of Apple's Personal Hotspot network is almost identical to the one on your iPhone.

WI-FI

You may connect to or leave a Wi-Fi network that you have saved.

Enable Or Disable Wi-Fi

- Choose the app for setting up.
- Choose Wi-Fi and then the switch for turning Wi-Fi on or off.

Quickly toggle the Wi-Fi.

From the screen's bottom, swipe up to get to the Control Center and choose Wi-Fi from the menu that appears.

Sign up for a Wi-Fi network

- When a device has its Wi-Fi turned on, it will immediately begin searching for available networks to join. Choose the Wi-Fi network from the list that appears. To connect to a Wi-Fi network, you will need to enter the password for your network and then click the Sign in button.
- A lock icon will be displayed on networks that are secure. Select "Other" from the drop-down menu if you want to connect to a Wi-Fi network that is unavailable to the general public. To continue, you must enter the correct information for your network name (SSID), security, and password (if applicable).
- When an iPhone is connected to any Wi-Fi network, the icon for that network's strength appears in the status bar.

Getting rid of a Wi-Fi network

- From the Wi-Fi panel, click on the information icon next to the network you want to use.
- Choose to Forget this network, and then choose to Forget to confirm

BLUETOOTH

Turning Bluetooth on/off quickly

Swipe up from the screen's bottom to access the Control Center, then tap the Bluetooth icon.

Bluetooth can be turned on or turned off.

- Choose the app for setting up.
- Tap Bluetooth. Use the power switch for turning the Bluetooth switch on or off

Connecting to a device

The gadget will start looking for other Bluetooth devices on its own. The available devices are displayed. Choose the device you want to connect to.

Disconnecting a device

- Click the info icon beside the device you want to use.
- Click on the option to disconnect.
- Click "Forget This Device" on the device's context menu to remove a device from memory. To confirm, select "Forget device" from the list.

5G CONNECTIVITY

5G connection is supported by the Qualcomm Snapdragon X65 modem included in the iPhone 14 Pro variants. In the United States, iPhone users may connect to mmWave and sub-6GHz networks. However, in other regions, only sub-6GHz networks are available.

How to switch from Wi-Fi to 4G/5G

1. Go to the **Settings app**
2. Tap on the **Cellular** option
3. Select **Cellular Data Options**
4. Tap on **Voice & Data**

5. The 5G Auto is the default. Select *5G On* to utilize it whenever available when you need faster speeds
6. Alternatively, tap *LTE* to turn off 5G and enhance battery life.

Chapter 6: iCloud Account, Apps Payment, How To Setup Credit Card Or PayPal

Apple ID and iCloud

You need an Apple ID to access Apple services like FaceTime, Apple Books, App Store, iMessage, iCloud and more.

Sign in with Apple ID
You can do this the first time you set up your device or after you have turned it on on your device for the first time.
- In the Settings app, click Sign in to your iPhone.
- Key in Apple ID details or choose to create an Apple ID account.

- Then enter the verification code if you turned on two-factor authentication.

Change Apple ID Settings
- In the Settings app on your iPhone, click on your name.
- Then choose the options accordingly to make your changes/updates.

Add Account Recovery Contacts for Apple ID

Add contacts that would help you reset your Apple ID profile if you ever forget your login details. To choose the contacts,
- In the Settings app on your iPhone, click your name.
- Select Password & Security and click the Account Recovery option.
- Click Add Recovery Contact to get prompts to choose the contacts.

Change your iCloud Settings

iCloud is a secure way to store your videos, backups, photos, documents, and more. It also ensures that they are automatically updated across all your synced devices.
- In the Settings app on the iPhone, click your name & choose iCloud.
- On the next screen, you will see your available storage. iCloud offers free 5GB storage, after which you would need to buy more storage.
- On this same screen, you can turn on the button for items you want to store on the cloud.

How to use iCloud

iCloud can automatically back up the iPhone.

Additionally, the following data may be saved in iCloud and synchronized across your iPhone and other Apple devices:

- Photos and movies
- Files and documents
- iCloud Mail
- Contacts, Calendars, Reminders, and Notes
- Information from third-party compatible applications and games
- Messages threads
- Passwords & payment methods
- Safari bookmarks and open tabs

- Settings for Stocks, News, and Weather
- Home & Health Records
- The voice memos
- Your map favorites

Moreover, you may accomplish the following:

- Share photographs and videos.
- iCloud Drive enables file and folder sharing.
- You may find a lost device or share its position with friends and family using Find My Device.

Apple Pay

You can shop at Target and McDonald's using Apple Pay on your iPhone.

Here is how it operates.

Paying using your phone is frequently quicker and faster than finding your credit card or counting cash. That's why people like mobile payment applications.

iPhone, iPad, Apple Watch, and Mac users utilize Apple Pay. Your iPhone or Apple Watch may make in-person and online purchases with a linked credit or debit card. iPads and Macs can purchase things from supported websites. Face ID, Touch ID, or PIN will deduct the money from your account.

Apple Pay needs appropriate hardware and software (it opens in a new window).

Setting up Apple Pay

Since you would virtually never go anywhere without your iPhone, the best place to start using Apple Pay is on the device you always carry with you: your iPhone. You merely need to be aware of how to put it together:

- Start up the Wallet program.
- Click on the plus (+) sign in the upper right corner to make it bigger.
- On the screen setting up Apple Pay, click "Continue" to move forward.
- Choose "Credit or Debit Card" or "Apple Card" from the drop-down menu as the type of card you wish to use with Apple Pay.
- This option is only available if you don't already have an Apple Card.
- You can add a new card using one of two methods: either scan the card's front with the iPhone camera or manually enter the card details into the iPhone (name and the card number). When you are finished, you can move on by clicking the "Next" button.
- After providing the requested information regarding the card's expiration date and security code, proceed by clicking the "Next" button.
- Before you put your name on the terms and conditions, be sure you've thoroughly read and glossed through them.
- Select the method of verification for your account (email, SMS, or phone), then input the verification code when prompted. After that, all you need to do is click the "Next" button.
- Clicking "Next" in your web browser will allow you to begin using Apple Pay as soon as the information you provided has been validated by your financial banking partner or credit card issuer.

Using Apple Pay

When you first turn on your iPhone to use Apple Pay, enter your credit card information. But you might have skipped this step because you were eager to begin using the new smartphone. If you make a mistake the first time, you can easily revert back and activate Apple Pay or add other payment cards you wish to use.

- Start up the Wallet program.
- Click on the plus sign to get started (located towards the screen's upper the upper right corner of the screen).
- When you get to step three, you'll be asked to enter your credit/debit card information or fill out an application to get the Apple Card.
- At this point in the process, the information for a credit/debit card will be put in. If you are in a city that supports this, you can also enter the information for your transportation card on this screen.
- After that, the next display is where you can start collecting cards. If you've already backed up your iPhone, you can use your CVC code to add cards that are already there from a previous backup. As a bonus, for first-timers in interacting with Apple Pay, you can link a different credit card or a different credit card altogether to your profile.
- For the Wallet app to be able to scan your credit card number, you will be asked to put the credit card into the frame.
- In addition, you can manually enter the credit card number, date of expiration, and CVV code (credit card verification value). Then, when Apple has confirmed the information with the card issuer, the card becomes automatically added to the wallet if all goes according to plan.
- Apple will send you an email message when the credit/debit card is ready for use with Apple Pay.

Using an iPhone to pay online

- Double-tap your iPhone's side button to pay online. After that, pay using Face ID or PIN.

- Double-tap Side to view your cards. You may pay with your Apple Watch in-store or online. Double-tap Side after tapping a card to purchase it.
- To purchase anything online with your Mac, use Apple Pay.

Chapter 7: Must-Have Social Apps And How To Set Them Up (Messages, Email, And Social Media)

Messaging on the iPhone

Aside from the general Messages option, Apple has the iMessage service exclusively for Apple users. This section shares tips on using the different features of this app.

Sign in to iMessage

This is the first step to using the iMessage app.

- In the Settings ⚙, click **Messages.**
- Then turn on **iMessage.**

Send a New Message/ Reply to a Message

To send or reply to a message,

- Open the app & press ✏ at your screen's top to draft a new message or click a conversation to reply to it.
- Enter the receiver's phone number if starting a new message.
- Enter the message in the message field. You can attach a photo or video to your message.
- Then press ⬆ to send the message or ✖ to cancel.
- Press ‹ to return to the message list.
- Press ⬇ to download a photo of the conversation.

Share Your Name & Photo

Store your name and photo and choose the receivers that can see this information.

- Click ⋯ in the Messages app & press **Edit Name & Photo.**
- To use a new profile image, press **Edit.**
- For a new name, press the name field & enter the new one.
- To share or not share your photo & name, toggle the **Nama & Photo Sharing** button on or off.
- To select specific people that can see your profile, scroll to **Share Automatically** & choose an option.

Respond to a Specific Message in a Conversation

To quote one message in a conversation,

- Double-tap the message & press ↩.
- Type your reply & press ⬆.

Pin/ Unpin a Conversation

Pin conversations with people you frequently contact to open them quickly. You can do this in two ways:

- Swipe right on the conversation you want to pin & press 📌. Press 📌 to unpin.
- Hold the conversation & move it to the beginning of the list. Drag it down to the end of your list to unpin.

Mentioning People in your Conversation

Mention individuals to attract their attention to that message.
- Open a conversation & start typing the person's name within the text box.
- Click the full name once it appears.

Reply to or Send an Audio Message

To send a new message,

- Open a conversation & hold the 🎙 button to record your message.
- Press ▶ to hear your message & press ⬆ to send if satisfied or ✖ to cancel.

To reply,

- Raise your iPhone 14 to play the received audio messages, and raise it again to record and send your response.

You can turn off the 'Raise to Listen' feature with the steps below:

- In the Settings app on your iPhone ⚙, click **Messages.**
- Then switch off **Raise to Listen.**

Receive Notification When You are Mentioned

The iPhone 14 has a vibration feature that alerts you to a new mention in a conversation. To turn it on,

- In the Settings app on your iPhone ⚙, click **Messages.**
- Then turn on **Notify Me.**

Always Keep Audio Message

By default, iPhone discards all audio messages 2 minutes after you play them; except, if you choose the **Keep** option. To prevent the iPhone from deleting audio messages,

- In the Settings app on your iPhone ⊚, click **Messages.**
- Click **Expire** & select **Never.**

Message or Attachment Options

When you have an attachment in your messages, you can download, save, share, or delete it.

- To save, share or print a given attachment, press it & select ⬆.
- To copy the attachment, press & hold it, then press **Copy.**
- To forward an attachment or a message, press & hold the item, press **More,** and tap ↪.
- To delete a message or attachment, press & hold it, select **More,** & tap 🗑.

Create Your Own Memoji

These are animations that mimic your expressions. They are cool ways to chat.

- Open a conversation, press 😀, & then click +.
- Click each feature & customize it as per your preference to bring your character to life.
- Once finished, press **Done.**
- You can delete, duplicate, or edit a Memoji – click 😀, click the Memoji, & press ••• .

Use Memojis in Conversations

You can add Memojis in your messages with the steps below:

- Open a conversation, press 😀 & click the Memoji you want to use.
- Write your message & press ⬆ to send.

Managing Message Notifications

To customize message notifications,

- In Settings app , click **Notifications** & select **Messages.**
- Then select the options you want.

Mute Notifications for a Conversation

Silence notifications for busy threads in Messages:

- Press & hold the conversation & select **Hide Alerts.**

Block A Specific Number or Contact

To block someone from sending you messages,

- Find a message from this sender, click their number or name, move down & select **Block This Caller.**

Viewing & Managing Blocked Contacts

- In the Settings app , click **Messages.**
- Click **Block Contact** to see & manage the list.

Delete a Message

A message is one dialogue box from an entire conversation thread. To delete a specific message or messages,

- Press the message bubble to get a list of options, then press **More.**
- Select other messages you wish to delete, & press .

Delete a Thread of Conversation

Delete all the messages in a thread.

- Open the Message app & swipe left on the thread/ conversation.
- Then click **Delete.**

Social Media apps

The internet brings many individuals closer to each other irrespective of geographical distance. Social media apps are here to connect people.

1. Instagram

You can share photos, multiple snippets as "Stories," short videos in form of Reels, and longer videos facilitated by IGTV and discover more new things every day. In addition, you can connect with other people through the direct messaging feature, and you can go Live or ensure you tune in to other users' live videos and attend events & workshops in real-time.

Pros of Instagram:

- It has an appealing visual interface.
- Compatible with iPhone.
- Combines videos, photos, short videos, & textual content.

2. Twitter

This is an excellent place to stay alert to current updates on global news and various topics. In addition, you can connect with other users globally and share, like, and retweet memes, thoughts, and stories.

Pros of Twitter:

- Being up to date with the latest trends.
- Has lots of information.
- You can easily connect with strangers, friends, and celebrities.
- Compatible with iPhone.

3. YouTube

Designed for watching videos in niches that interest you ranging from travel, life hacks, technology, science, blogging, and more. You can get helpful content from creators like chefs, makeup artists, musicians, dancers, tech geeks, astrologers, gamers, and more.

Other social media apps include:

- **Clubhouse** - For audio-only networking concept/format of content.

- **Facebook** - A great social networking app with various features and settings.
- **LinkedIn** - An excellent social network that can transform your career. You can search for jobs, connect with recruiters, get business contacts, and build your professional network.
- **Snapchat** - A good app for connecting and communicating through fun visual content.
- **TikTok** – The social app that revolutionized the online world with short-form video format.
- **WhatsApp Messenger** – Great for sending messages, photos, videos, documents, and voice notes. Also supports audio and video calls.
- **Telegram** – A messaging alternative for WhatsApp. Used for sending media and files without limiting their type and size. It has seamless cloud storage, saving you more disk space.

Chapter 8: Detailed Instructions To Set Up The Camera And How To Capture The Best Photos/Videos

Capturing Photos

Learn to take pictures with your device's camera. Select from different camera modes like photos, videos, cinematography, pan, and portrait, and Zoom out or in to frame the shot.

Launch The Camera

To access the camera 📷 on your device, either slide leftward on the Lock screen.

Change Camera Modes

You will be sent straight to photo mode when you start the camera application. Swipe to the left or right to select between the various camera modes, which are as follows:
- Video
- Timelapse
- Slow-mo: Record a slow-motion video.
- Panorama
- Portrait
- Cinematic
- Square: Touch the Camera Cut button⊙, then touch 4:3 to pick from 16:9, 4:3, or square aspect ratios.

Zoom In or Out

To zoom in or out, press the 0.5x, 1x, 2x, 2.5x, and 3x buttons. To navigate more precisely, you should first press and keep your finger on the zoom control, then move the slider to the right or left.

Take Macro Pictures

You can capture a macro image or video using the Ultra-Wide camera on the iPhone 13 Pro and iPhone 13 Pro Max. Both phones have the capability. To do this, launch the Camera application, go close to the subject, up to 2cm close, and your device's camera will achieve focus automatically.

Take A Photo Or Video

Touch the Shutter icon to take a picture, or press any of the Volume keys.

Switch Flash On Or Off

Your device's camera will use flash automatically in dim settings. You can manually control flash before taking a picture by touching the Flash button ⚡ to switch it on or off. For example, click the Camera Control button ⊙, and then click the Flash button ⚡ under Frame to select Auto, On, or Off.

Take Pictures With Filter

Use filters to give your photos a color effect.

- Select Portrait or Photo, touch the camera control button ◉, and then touch the Filter button ⊛.

- Under the viewer, swipe to preview the filters; touch one to use.

Utilize A Grid To Straighten Your Shot

Enter the Settings application> Camera, and open the Grid to display a tab that can help you straighten and organize your photos on the camera screen.

Once you've taken a photo, use editing tools in the Photos application to align the photos.

Use Photographic Styles

On the iPhone 14, you can use Photo Style to personalize how the camera takes pictures. First, pick from the preset styles: Cool, warm, vibrant, or rich contrast; later, adjust the warmth and tone.

1. Launch the camera application, then press the Camera Control button ◉.
2. Touch the Photographic Style button ◈, and then swipe left to view the different styles:
    - Rich contrasts
    - Vibrant
    - Warm
    - Cool

To change the Photo Style, use the Tone & Warmth controls at the frame's bottom and move the slider left or right. Then, touch Reset images ↺ to reset the values.

3. Touch the Photo Styles button ◈ to use the style.

Take Live Pictures

A Live Picture captures what takes place before & after the picture is taken, as well as the sound.

1. Launch the camera app in Photo mode.

2. Ensure that Live Photo is activated. When it is active, you will see a live picture button ⊚ directly on top of your camera. Touch the live photo button to activate or deactivate Live Picture ⊚.
3. Click on the Shutter icon to capture the live picture.
4. To play the live picture, touch the lower part of the picture thumbnail in display, then long touch your display to play.

Note: Live Pictures are not available when ProRAW is active

Take Pictures With Burst Mode

Burst mode is good when taking pictures of a moving object/subject. This mode takes pictures at a high speed, meaning frame-by-frame shots can be picked.

1. You can take burst mode pictures by swiping the shutter icon to the left to capture fast pictures.
2. The counter displays the number of pictures taken.
3. Stop the touch input by lifting your finger to stop taking pictures.
4. Touch the Burst thumbnail to choose the pictures you plan on saving, and press Select.
5. To save a picture, click the Done button in the lower right corner of the image.
6. To delete the whole picture, touch the thumbnail, then touch Delete 🗑.

Take A Selfie

1. Navigate to the front camera by pressing ⊚ or .
2. Keep your iPhone in front of you.
3. Press the Shutter icon.

Adjust The Volume Of The Shutter

1. Use your device's volume buttons to change the shutter's sound volume.
2. Otherwise, open the control center. Drag sound slider when the Camera application is on.

3. Flip the mute switch on the side of your device.

Recording Videos

1. Select the video mode.
2. Tap the record icon. You can do any of the below while recording:
   - Press the white shutter button when you need to take a picture.
   - Pinch your display for zooming capabilities.
   - Hold the 1x, and slide the slider to the left for more accurate zooming.
3. Touch the record icon to end the record session.

Record A Cinematic Video

With cinematic mode, your phone detects the subject of the video and puts them in focus during the recording; if another subject is detected, your Phone will shift the focal point. You can adjust the focal point manually or change it in the Photo application.

- Select Cinematic mode.

On the iPhone 13 Pro & Pro Max, you can press 1x to zoom in before you start recording.
To adjust the depth effect, click the Adjust Depth button⊙ and then slide the slider left or right before you start to record.

- Tap the record icon to begin recording.
- The yellow frame on your display shows the individual is focused on; the gray frame shows that the camera has detected someone, but the person is not being focused upon. Touch the gray frame to switch the focus, and touch again to lock the focus on that individual.
- If there is nobody in the video, touch anywhere on your display to determine the focus point.
- Tap the record icon to end the recording session.

Record a Quick-Take video

1. In Photos mode, hold down the Shutter icon to record a Quick-Take video.

2. Drag the Shutter to the right and allow it to go over the lock so that it can record manually.
3. Tap the Record icon to end the recording session.

Recording Slow-Mo Video

- Select the Slow-Mo mode. Then, touch the flip camera button to record with the front camera in Slow-Mo mode.
- Touch the Record icon to start recording.
- Press the record button again to end the record session.

To play only a segment of the clip in slow motion while keeping the remainder of the video playing at a steady pace, hold down the video thumbnail while pressing the Edit button. The diagonal lines under the viewer can be shifted to select the portion of the video you want to slow down.

Share Photos And Videos

You may use the Photographs app to exchange photos and videos with other apps you install, such as Messages or Mail. The photos app even suggests the best shots from a given event and recommends individuals to share with.

You can share a single photo/video or multiple items via messages, email, airdrop, or Bluetooth among others.

Scan QR Code

You can use your camera to scan a Quick Response (QR) code for a link to a site, an application, a ticket, a coupon, etc. The camera detects and displays the QR code.

- Open the camera, then position your Phone so that the code shows on your display.
- Touch the notification on your display to be directed to the appropriate site or application.

Chapter 9: Health and Fitness

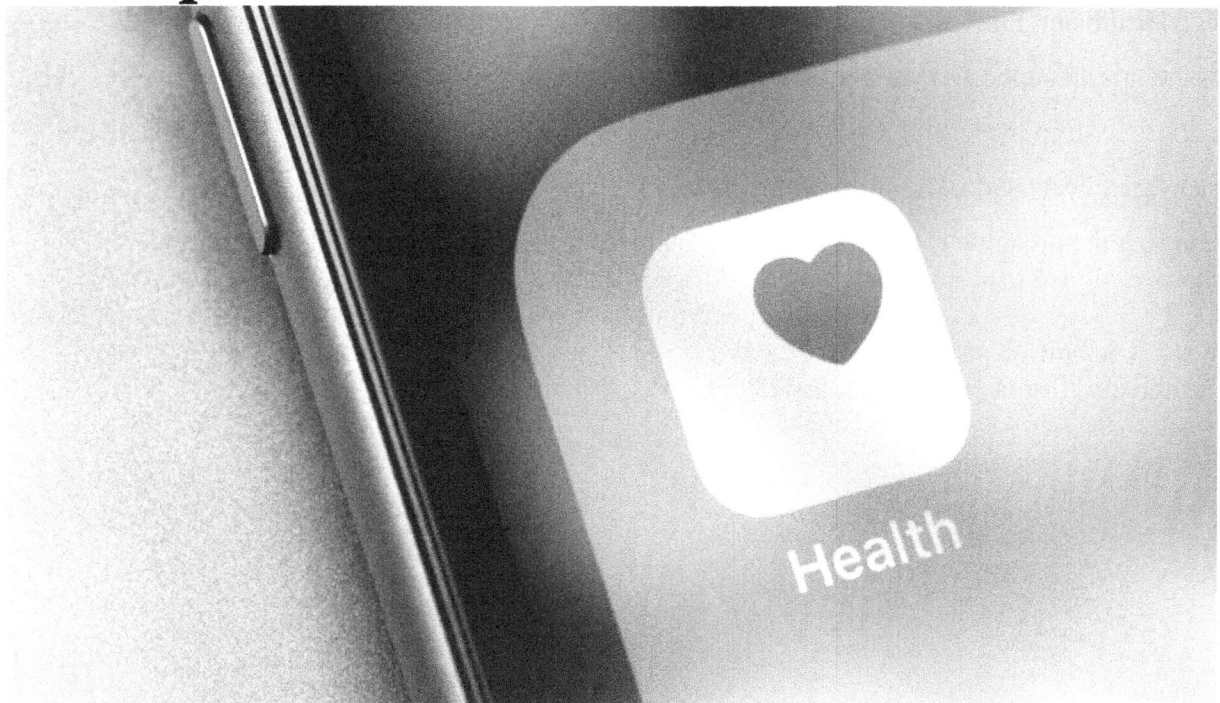

When it comes to your phone, the apps you use are essential. The Health app tracks your steps, workout sessions, weight data, and more. The Activity app tracks how active you are throughout the day—running or simply getting off the couch. You may also share this information with your friends and family to keep them motivated!

The iPhone 14's Health and Fitness program tracks your activity level and uses machine learning to determine which metrics matter the most. Learn how this program works and how you can get the most out of it. We'll also cover how to customize the program to suit your personal preferences, but before we get started, let's go over the basics.

The iPhone 14's Health And Fitness Program

The iPhone 14's Health and Fitness program has a few considerable benefits. The device's display is a standard rectangle with rounded corners, and it measures 6.06 inches diagonally.

This is smaller than the actual viewable area, but it is still large enough for people to see it clearly even when in portrait orientation.

The Health and Fitness program integrates third-party apps and health metrics into a single dashboard. It supports a variety of activity and health apps, including apps that track your meals, activities, sleep cycle, and meditation practice. You can also use third-party apps to track your menstrual cycle and track other health metrics.

The Health app on your iPhone can track your activity level and store it for a later review. The app is designed to securely store data from your iPhone, Apple Watch, and other compatible devices. It also allows you to manually input data, such as your body measurements or menstrual cycle. The built-in sensors in the iPhone can also track your steps and walking speed. In addition, the app also displays highlights and details of your activities.

Monitor your walking pace via the Health app on your iPhone

When you carry your iPhone in a pocket or holster around your waist, the Health app assesses your balance, strength, and gait using unique algorithms. You may get a notice if your steadiness drops or remains low, and the notification can be instantly shared with a close friend. The Health app may also recommend workouts to enhance your walking balance.

Prioritize your Health with the Health App

For the most part, the Health app does its work in the background while you use other tools. Once you create a profile, you can set up most tracking apps to automatically share your data with Apple Health.

When you open the **Health app**, you'll first see a summary page that shows things like steps taken, calories burned, recent activities, and more. You will also immediately see some trends and totals.

For example, you can see a comparison of the number of steps taken compared to previous days, the duration of training sessions, the average heart rate, and the like. This summary is tailor-made for you based on machine learning algorithms.

By clicking on the **Overview** tab at the bottom of the screen, you can view additional information in a particular category. For example, you can analyze data about your sleep, heart rate and respiration rate, and more. Any of these expanded tabs will also give you the option to see the data sources. This way, you can see exactly where Apple Health is getting information from.

The Health app stands out for the depth and wholesomeness of the information it offers. Since the application works with so many platforms, you see data on your smartphone that is not available elsewhere, including information related to hearing (shouldn't it be time to take off your headphones?), mindfulness, and even handwashing.

You can also provide additional information about yourself by clicking on the profile icon in the upper right corner and selecting **"Health Information."** This is useful in case you are in a medical emergency.

Track Your Steps

Carry your iPhone with you to keep track of how many steps you take, how many flights of stairs you climb every day, and how well you walk. Start the Health app, and then click Summary to see how many steps you've trodden. Scroll down to set up Walking Steadiness Notifications. Tap Edit to add more activities to the Summary screen, such as Flights Climbed or Walking + Running Distance.

Track your health

The Health app uses your iPhone to keep track of your health and fitness data, like how many steps you take, your resting heart rate, and how much sleep you get. Trends let you see how certain indicators change over time, and you can set up alerts to let you know when new trends are recognized. Tap Summary in the Health app, and then scroll down to Trends

The Health App

The iPhone's Health app is quite simple. It starts with your weight, then tracks information about your activities and diet daily.

The Health app is also where you'll find Fitbit sync and third-party apps that integrate with the Health app.

Steps: What's the total number of steps you've taken today? This feature will appear in the Health app when your iPhone's clock hits midnight. The basic idea is to track your step count every day, either manually or by syncing to an app like the Fitbit Surge, Jawbone Up2, Nike + Fuelband SE, or Apple Watch (more info below).

A tip for measuring your step count accurately is to pick a starting point, then walk in a clockwise direction for at least 100 steps. Then, multiply that number by 4, and you should be able to get an idea of how many steps you take in a mile!

Workouts: There are a variety of workouts to suit every fitness level. Pushups, sit-ups, cardio activities, and even different sorts of running are all options! The only stipulation is that your iPhone has a heart rate monitor and a gyroscope, so the health app can accurately track how many calories you burn throughout each workout.

Some other cool features come with this app. For example, you can record how many calories you've burned while weightlifting or running. This is great to see if you need to go over the top on your workouts to gain muscle size or if you need some extra motivation before hitting the gym!

Heart Rate: How many times in each minute were your heartbeats measured? With this data, the health app will show you where and when the best workout zones are on your heart rate chart.

Health + Health Kit: The Health app is the star of the cool features on your iPhone. However, the Health app is useless without the Health Kit. This feature gives you access to other fitness apps and data that you can use to get in better shape. In addition, you have control over how much information these apps have access to. The Fitness & Health app greatly explains the process and how to get started!

Food: If you're already tracking your food intake, the health app will now sync those calories with your individual Fitbit account linked to the health app. This is a fantastic way to track how your food affects your body. The health app also gives you a calorie counter for any food with nutritional information.

Strava: In addition to steps, heart rate, and calories burned, the health app also displays information of all kinds on Strava using its tracking platform. You can also sync your other apps with Strava through Health Kit. It's a great way to track how much time and energy you're putting into your workouts!

The Activity App

The activity app is simple compared to the health app: it gathers daily data about how active you are. This includes standing, walking, running, and cycling sessions. The greater your degree of achievement, the more active you are each day—up to "Exercise Ring" if you get 30 minutes of moderate exercise every day.

Inactive Minutes: I'm sure you've all heard of inactivity. One of the most frequent fitness issues in America is that nearly half of Americans are not as active as they should be daily. The more inactive you are, the less likely you are to lose weight or curb your obesity problem. With this feature, the iPhone will try and motivate you by counting how inactive you are each day!

Activity Rings: The "Exercise Ring" for your activity is earned when getting 30 minutes of activity every day. You can also earn other rings if you're more active throughout the day or when you hit the goal of playing a game.

Calories Burned: This is where the iPhone will show you how many calories your workout sessions burn during a day to help you decide how much food to eat to maintain your weight and achieve a better body!

Other apps that can sync with Apple Watch: Strava, Runtastic Running, Moves, 7 Minute Workout, MyFitnessPal, Nike + Fuelband SE, and more! If you have one of these apps connected with Health Kit, it'll automatically sync exercise data all day long for free. Otherwise, it'll only sync for about 20 minutes at a time.

Health Records And Fitness Data

Create an Emergency Medical ID account.

If you have health insurance or a health plan requiring a medical ID, you can also sign up for that free service. This will allow you to store your medical information and access it from any iPhone or Apple Watch.

After your iPhone sends you a text message with an invite, you'll be able to join up for this service through the Apple Health app. (You'll need to verify your phone number.) You'll then receive the item in the mail with instructions on how to go about creating your account.

Once you've created your account, you can go to the Health app and enter your medical information. This is an excellent alternative for folks with health problems that need to be addressed immediately.

Updating Your Health Profile Manually

If you want to update your health profile manually, you can head to the Health app. First, you'll see a tab for Health Records, and this will be where you'll be able to see each of the different information pieces we talked about in the previous section: allergies, blood pressure, weight, glucose, and more.

Tap on any one of these categories, such as Blood Pressure or Weight. This will open all sorts of data points for that category. If a point is incorrect (like if your scale doesn't show the correct weight), tap on that point and it will allow you to change or correct it.

You can also go to the Settings screen and the Manage Health Records section. Here you can update any recommendations from your doctor stored in your Health app. (These are helpful if you want to track a health condition you're currently dealing with.) You can also share this information with your doctor by going into the Notes section by selecting Share Notes > Medical App Notes.

Creating a New Health Record

The next stage is to make a new health record to keep track of things like weight, blood pressure, and stress levels. To do so, go to the Health app and select Add Record > Keyline Medical Record.

Name your health record, tap on Save and then tap on Done.

Your new health record can be found under Keyline Medical Record in the Health app. By default, there will be just one record. If you want to add more, tap on Add Record and follow the instructions to add a new record to your keyline medical records. In addition, you can include or exclude data points you don't want to be updated by going into the Edit option under Keyline Medical Record > Edit Keyline Medical Record > Exclusions.

Disabling Health Data Sharing

By default, the Health app will automatically share your information with Apple if you've enabled medical alerts or any other data-sharing options within the app itself. You can disable this at any time.

To do so:

Return to the Settings app and select Privacy > Health from the drop-down menu.

You'll find an option for Sharing with App Developers if you scroll down. Depending on your needs, toggle it on or off. For instance, toggle off if you wish to disable sharing any health data. You'll still be able to view your information in Apple Health and any medical app as long as this is turned on, but other apps will not access them if you turn it off.

You can also go back into the Keyline Medical Record within the Health app and change your sharing settings.

Tracking Your Fitness Data

The Health app also keeps track of your fitness activities, and you can use this to keep an up-to-date log of everything you're doing daily. This information allows you to set yourself up for success by helping you know when to exercise or take certain medications based on specific activity and sleep levels.

To start tracking your fitness information, go to the Health app and tap the Activity tab. Your overall activity levels are broken down by exercise and general activity throughout the day. On this page, you can also toggle on or off any of these categories, such as Standing Hours or Exercise Minutes, to track only some of these activity levels. Or, you can tap on each category to show more detailed information about what you did during that period.

Notes:
- If you're using a wheelchair or don't exercise regularly, remove all of these options from this page by going into Edit > Exclusions>Exercise Minutes and turning them all off.
- If you're a runner and don't want the activity splitting up your training into different exercise segments, then stop recording this information by going into Edit > Exclusions and turning those options off.

Document All of Your Health Information

Many people with chronic illnesses or physical limitations use the Health app for tracking their health history and various health conditions. This allows them to record everything from pain and physical activity levels to changes in medication and vitamin intake. If you want to take this a step further, the Health app allows you to document your health information over time.

This is handy if you want a visual representation of your changes throughout the years or if you'd like to share this with a doctor for some reason. To create a new entry in your health record, go into the Health app > History > New Entry. Here, you'll be able to select from one of your previous records and make all the changes that have occurred over time.

How To Set Up Sleep Timers

The sleep timer is a feature on the iPhone that allows you to set a specific amount of time for your device to go to sleep. This can be useful if you want to make sure your device doesn't stay awake all night or if you want to save battery life by having it turn off automatically after a certain period of inactivity.

To use the sleep timer, open the **Clock app** and tap on the **Timer tab.**

Then, select how long you want your device to sleep from the options at the bottom of the screen.

Once you've made your selection, tap on **Start** in the upper right corner and then put your device down somewhere where it won't be disturbed.

The screen will dim after a few seconds and then turn off completely when the Sleep Timer expires.

Chapter 10: How To Use Dynamic Island (14 Pro And Pro Max Only)

There is no notch on the iPhone 14 Pro and Pro Max. This is a departure from the iPhone X, which was released in 2017. Instead, Apple has updated the TrueDepth camera's hardware. The proximity sensor has been relocated beneath the screen and the dot projector, camera, and other elements have smaller cutouts.

The right side of the display features a small circular cutout for the camera. The left side has a pill-shaped cutout for Face ID's TrueDepth system, which Apple is combining with software into a single cutout that is pill-shaped we refer to as "Dynamic Island."

By incorporating it into the interface, Apple's Dynamic Island is no longer a simple cutout. Instead, it changes shape depending on what is displayed on the screen and serves as an information hub in the center.

Apple uses Dynamic Island to access iPhone apps and services quickly. Dynamic Island extends to mimic the Face ID confirmation interface while making an Apple Pay payment and when on a phone call so that you can use the phone's controls.

It shows Apple Maps directions, timers, Apple Music, and AirPods connectivity. In addition, it integrates with Live Activities in iOS 16, allowing you to check sports scores, Uber rides, and other information from the top of your iPhone screen without leaving the app you're in.

It can fold down so you can see what's on your screen or open up to interact with what's on the display. In addition, Dynamic Island includes a feature that allows it to be divided into two separate cutouts that display different information. Third-party apps can also add Dynamic Island support.

Chapter 11: How To Setup Always-On Display

The Always-On display of the iPhone 14 series lets it dim the Lock Screen while you can still see essential details like the widgets, clock, and wallpaper. It does this using new technologies that make the display very power efficient.

This is fantastic if you only wish to have a quick check at the time or read important data from the widgets on the Lock Screen without having to unlock the iPhone and open a given app.

The display is constantly active, but to preserve battery life, its refresh rate and brightness are automatically reduced to as low as 1Hz without affecting the display's function. The iPhone 14 has an always-on display standard, and both versions have it turned on by default. **The following actions should be followed to either switch on or off the always-on display:**

- Select Settings > Display & Brightness from the menu.

- Navigate to the option labeled "Always-On" to activate or deactivate the display that stays on all the time.

Chapter 12: How To Setup Widgets

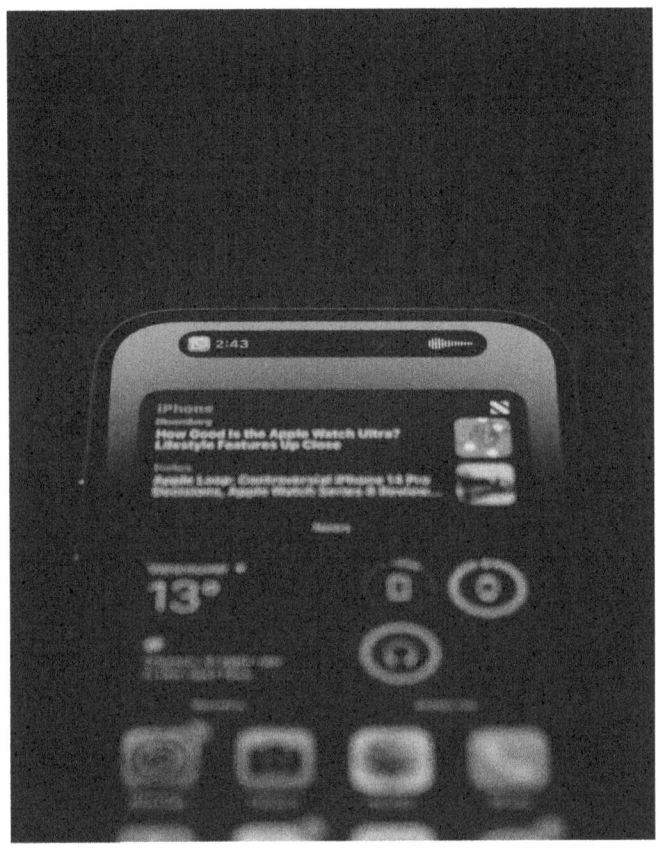

You can use widgets on the Home Screen with iOS 16 to keep your favorite information close at hand. Alternatively, you can swipe right from the Lock Screen or Home Screen to utilize widgets from Today View.

Add Widgets to your Home Screen

Widgets show you different information from different apps and services on your phone. The widgets are found in iPhone's Today View, but it is possible to add them to your home screen to access the information quickly.

- Find the position on the home screen to insert the widget, and press & hold the background of the screen until the apps start jiggling.

- Press ✛ on your screen to open the iPhone 14's widget gallery.
- Search for a widget, click the one you want, then move right & left to see the different sizes. Each size displays different information.
- Click **Add Widget** when you find the size you like.
- Then drag the widget to any part of the screen & press **Done.**

View Widgets in Today View

To view widgets in this view,

- Go to Home or Lock screen. From the left edge of either screen, swipe right & then scroll down & up.

The widget on the lock screen must be turned off if you cannot access them. To turn it on,

- In the Settings app on your iPhone ⚙, click **Face ID & Passcode.**
- Key in your passcode when a prompt appears & switch on the **Today View & Search** button.

Edit a Widget on the Home Screen

By default, each widget has some specific information it displays, but you can choose the information you want to see from a widget with the steps below:

- Return to the home screen & find the widget.
- Press & hold the widget until you see the quick action menu.
- Click **Edit Widget** or **Edit Stack** & choose your options.
- Tap **Done** to finish.

Remove Widgets from Home Screen

- Press & hold the desired widget until you see the quick action menu.
- Select **Remove Stack** or **Remove Widget**, before pressing **Remove.**

Moving Apps & Widgets Around on your Home Screen

It is possible to change the location of widgets as per your tastes.

- Return to the home screen & find the widget.

- Press & hold the widget until you see the quick action menu.
- Click **Edit Home Screen** & drag the widget to a different spot on the same page. To shift it to a different home screen page, drag the widget to the right edge of your screen & wait a few seconds for the next page to appear. Hold until you get to the page you want, then drop/release the widget.
- Press **Done** to finish.

Chapter 13: iPhone Accessories

What Comes In The Box

The iPhone's Box currently comes with the following accessories:

Lightning-USB cable: Connect to a PC/computer and an AC adapter.

USB power adapter: Meant to connect and charge the battery. The size and type of adapter depend on the country or region.

You can also use the charging cable for connecting your iPhone to a PC/computer to download and transfer files, use your iPhone as a second screen on your Mac, and more.

What You'll Need To Buy

The following accessories can easily be obtained from Amazon or other electronics shops. Although you can use your phone owning them, your user experience will improve significantly upon having them. Also, you'll be able to use your phone for a long time without spending money to repair it or get another one.

1. **Phone Case**: For some people, a phone case is a must. Phones can sometimes be slippery. With a protective case, your phone will be protected, even if you accidentally drop it. Various types are available (silicone cases, leather, etc.), and they are quite affordable.

2. **Screen Protector**: This is another accessory you should prioritize getting. Although you may not see the need immediately, you will over time. A screen protector/guard can prevent your phone's screen from scratches by keys and other sharp objects. But more importantly, they can prevent the cracking of the phone screen if you drop it, which will be a disaster if it happens.

3. **Wireless Charger**: For those who don't like having a lot of cords and wires around, you can get a wireless charger for your phone. Some phone cases (batteries) have inbuilt wireless chargers, so try to do a little search (you'll save money and get more done).

4. **Headphones**: The most popular headphones for iPhones are the Apple AirPods. They don't have troublesome wires and connect to your iPhone superfast. So, enjoy listening to songs, making phone calls on the go, and doing other crazy stuff with a cord-free experience.

5. **Battery Pack (or Power Bank)**: There are various brands in the market (e.g., CONXWAN, Attom Tech, etc.). Although these are optional, they are instrumental, especially if you are on a business trip, traveling for leisure, or living in an area with a poor power supply. A power bank is used to store power that you can use to charge your phone whenever needed. Many of them can juice up your battery multiple times (sometimes up to 8). Moreover, you can easily carry them around stress-free due to their portable sizes.

6. **PopSockets**: These are some trendy accessories among iPhone users. They are used for holding your iPhone in one hand while you take photos, browse the internet, and more. You could also use them as a stand for your phone. Due to their popularity, you may

already have one. If you don't own one yet, know you'll be making a good investment if you get one.

Chapter 14: Apple Car Play, Music, Text To Speech, Reading, Etc.

CarPlay

If you have an iPhone in hand, a vehicle with a fancy screen, and a direct connection for your phone, all you need to do is drive. Apple CarPlay can automate your driving experience by displaying your podcasts, music apps, calendar, contacts, and other programs like maps on the center console display.

Apple CarPlay is a wonderful feature for iPhone owners. It gives you a completely hands-free experience while driving and allows you to send and receive text messages and use applications while driving. However, if you've recently purchased an iPhone 14 Pro, you might be having trouble using CarPlay. If so, you've come to the right place.

Apple CarPlay has been introduced as the next evolution of the iPhone experience in cars. It brings Apple iOS applications to the dashboard and instrument cluster, allowing users to control vehicle functions through the device. It also offers customization options, such as curated gauge cluster designs, widgets, and more.

Check for compatibility

You must first determine if CarPlay is supported by your vehicle and phone.
CarPlay is compatible with over 600 vehicle types. Older models will require third-party infotainment systems to enable the connection between the phone and your car. All manufacturers—including Chevrolet, Honda, and Jaguar—have compatible vehicles as old as 2016. Unfortunately, CarPlay is not compatible with Tesla yet.

Do you need a cord?

All your systems are now operational. You have your phone and are seated in the vehicle. But for your vehicle to talk to your phone, you probably need a cable.
Only a small number of (expensive) vehicles have specific wireless CarPlay connectivity, including select Mercedes, BMW, and Audi models. Therefore, you will need a USB cable to connect to most automobiles. There is probably a USB port in the vehicle's center console area. The two alternatives operate as follows:

Pairing up

Get your phone ready with a suitable vehicle and prep to have the CarPlay connect wirelessly through Bluetooth. Add the vehicle as a recognized device to ensure the car pops up automatically among the My Cars list. Check **"Settings" > "General" > "CarPlay."**
Turn the Bluetooth toggle on and off, then check for the automobile to link before adding it for the first time. Check to the bottom for **"Other Devices"** to initialize it. Pick the vehicle, then adhere to any matching instructions that appear.

Connect it

It's really simple when using a USB cord. The CarPlay interface should launch on the vehicle's screen once the phone is plugged in with an unlocked home screen. While it is linked to the automobile, your phone can only be used to a limited extent on a handset. You may be limited to utilizing the map on the screen in the automobile, for instance, while using a GPS app to navigate.

Connect while driving

Once connected, you get a home screen with rows of squares representing all the accessible applications that resemble an expanded home screen version of the iPhone.
Additionally, instead of having a solid background, CarPlay now allows wallpaper displays on iOS 16.
You will notice more applications that support CarPlay in addition to built-in Apple apps such as Contacts, Music, Podcasts, Messages, Audiobooks, and Maps. This covers other mapping services like Google Maps. To utilize applications in the automobile, you must first download them to your phone.

Changing CarPlay

You can sometimes exit the CarPlay panel and return to the default screen of the console. If you didn't disconnect anything, you're still linked to CarPlay. All you have to do to return to your applications is locate the menu item or CarPlay icon.

Turning off

Once you arrive at your designated destination, you should disconnect the phone. Your phone resumes its normal operations as soon as you power the vehicle off. You can also unplug your USB cable.

Dictating Text-To-Speech On Your iPhone

If you're looking for a way to make your Apple iPhone 14 more useful, you may have heard of Text-to-Speech accessibility. Text-to-speech tools can convert selected text to audio. These tools are available in the Settings menu under Accessibility and Spoken Content. They can even insert question marks and periods for you.

The first step is to enable the text-to-speech feature on your iPhone. To do this, you'll need to open the Settings app and scroll down to Accessibility. In the Accessibility section, tap the Speak Selection option. You can also adjust the speaking rate by adjusting the Speaking Rate slider.

Another notable feature of the new iPhone 14 is that it no longer needs a physical SIM card. Instead, you can use the electronic SIM, which has been around since the iPhone XS. While Apple has eliminated the physical SIM card slot from its US lineup, users in the UK can continue to use a physical SIM card.

Not all languages, nations, or areas may offer dictation, and the features may change. Charges for mobile data usage may apply.

Activate dictation

- To access the keyboard, head to Settings > General.
- Switch on the dictation feature.

Dictating text

Wherever you wish to insert text, tap to set the insertion point.

Within the text area or on your on-screen keyboard where it may appear, press the Dictate key (for instance, like in Messages). Then start talking.

- Ensure Enable Dictation is turned on in Settings > General > Keyboard if you can't see the Dictate key.
- The iPhone automatically adds punctuation while you continue speaking to insert or add text. In addition, you can enter an emoji by stating their names (such as "happy emoji").
- To disable this feature, check on Settings > General > Keyboard and deactivate Auto-Punctuation.
- Select the Stop Dictation button when you're done.

Chapter 15: Best Tips And Tricks

Have you considered how you could make the most of your iPhone? Here is a collection of tips and tricks to assist you with mastering the technology.

1. **Action mode lets you get a steadier video.**

With iPhone 14 Pro and iPhone 14 Pro's Action mode, you can capture smooth videos even when moving around a lot.

- In the Camera app, swipe to Video mode.
- By tapping the button, you can activate action mode.
- To begin recording, press the "Shutter" button.

Action mode works best in well-lit conditions, and if there isn't enough, the camera will prompt, "More light required." You can configure your camera to use Action mode when there is less light.

- Tap Camera in the Setting app.
- Tap Record Video.
- Activate Action Mode Dim the light.
- Action mode can record up to 60 frames per second of either 1080p or 2.8k video. It can play Apple ProRes video formats or Dolby Vision HDR video formats on iPhone 14 Pro models

2. **Use your iPhone 14 Pro to set up crash detection**

If you don't cancel the detection, your iPhone 14 Pro will automatically make an emergency phone call after 20 seconds if it detects a serious car accident. If you become unresponsive, the iPhone plays an audio message for emergency services telling them that you were involved in a serious crash and giving them your GPS coordinates and a rough search radius.

When a crash is found, Crash Detection won't cancel out any other emergency calls that have already been made.

Crash Detection is already turned on. Settings > Emergency SOS > Call After Severe Crash lets you turn off alerts and automatic emergency calls from Apple after a serious car accident. They will still be told if they have set up third-party apps to be notified when their device crashes.

3. **Use Satellite Emergency SOS**

With this iPhone 14 trick, you can text emergency services using Emergency SOS via satellite when you are not on a cellular network or Wi-Fi.

The use of Emergency SOS through satellite helps one get in touch with emergency services in rare cases when no other alternatives exist. For example, if you try to call or text emergency services and you can't get through due to the phone being out of network reach or Wi-Fi coverage, your iPhone will try to connect you to help services through satellite.

Here are steps for connecting an iPhone to a satellite, so you can call 911 if you need to:

- Hold the phone in your hand in a natural way. No need of raising your arm or phone. Don't place the phone in a backpack or pocket.
- Make sure you're outside and can see the sky and horizon.
- Remember that trees with thin leaves could slow down the connection, and trees with thick leaves could block it. The connection can also be blocked by hills, mountains, canyons, and tall buildings.
- If there is need to turn right or left or move to get around a signal block, your iPhone will tell you what to do. Just follow the instructions on the screen.

To send an emergency message by satellite

First, try to get help by calling 911. You might be able to make the call even if your regular cell phone network isn't working.

If your call doesn't go through, you can send a satellite text to emergency services:
- Tap the Satellite Emergency Text button.
- You can also text 911 or SOS by going to Messages and tapping Emergency Services.
- Tap Call for Help.
- Answer emergency questions with simple taps which will describe the situation you are in.
- You may also alert the emergency contacts that you have called the police or fire department, where you are, and what the emergency is.
- Connect to a satellite by following the on-screen instructions.
- Once connected, follow instructions to maintain the connection while you send the message to emergency services.

- Upon successful connection, your iPhone initiates a text conversation with emergency personnel. It shares critical information such as your location particulars, Medical ID & emergency contacts, answers to emergency questionnaire, and how much battery life the iPhone has left.

4. **Recover recently deleted Message chats**

If you accidentally delete messages from the Messages app and need them later, you can recover them on your iPhone 14 Pro. When you delete a message, it is moved to the Message app's "Recently Deleted" folder. Here's how to access the "Recently Deleted" folder in Message.

- Launch the Messages app.
- Press the Edit button located in the upper left corner.
- Choose "Recently Deleted" folder.
- Choose the chat from which you wish to receive messages.

Note: Deleted messages will stay in the "Recently Deleted" folder for a period of 30 days. After that, they will be gone forever.

5. **Deleting or removing contacts**

In older versions of iOS, it was lengthy to delete a contact from an iPhone. You had to go to the contact's details, click the Edit button, scroll down, and then click the Delete Contact button. Apple has finally listened to its customers because iOS 16 makes it easier to delete a contact. To remove a contact from your list, you can now tap and hold on to it and click the prompt to delete the contact.

6. **Turn off "Lock to End Call."**

One thing that iPhone users don't like is that pressing the lock button ends the call. This is especially true for people who switched from Android. With this iPhone 14 trick, you can finally turn off "Lock to End Call."

- Open the app for Settings.
- Check out Accessibility.
- Choose Touch.
- To end the call, turn the Lock off.

7. Use Siri to end a call

To end a call on an iPhone 14, you can tap the end call button on the screen or the side button. You can also have your virtual assistant, Siri, end the call for you in a very bossy way. For example, say, "Hey Siri, hang up" while you are on a call, and the assistant will do it for you.

This is a convenient and hands-free way to end a call, but you should be careful because the person on the other end will be able to hear what you tell Siri.

8. Use Siri to restart your iPhone 14

You can also restart your iPhone 14 with Siri, another secret trick. Just say "Hey Siri" to wake up your assistant and say "reboot" or "restart." When Siri asks for confirmation, all you have to do is say "Yes" or "No." Siri lets you do the action without unlocking your device so that it can be done hands-free.

- Find new apps for your iPhone. Using an app like the App Store, which has produced a complete library of every app accessible on the market, is a terrific method to locate new apps for your phone.
- Vibrations from emails can now be set up on your phone so that they vibrate for specific events, such as when you have received an email with exciting news or information about something that interests you.

- If you want to access multiple calendars on your iPhone, go to Settings, then choose Mail, Contact, and Calendars. You'll see a new option for Calendars here, where you can import calendars from other sources.

- If you don't want your friends or family members to know where you are at any given time, change the Location Services setting and set it to Off. This way, the iPhone will not reveal your location if the people you are calling do not have GPS on their phone or if the phone is turned off.

- Are you a social network fanatic? Are there specific Facebook updates or Tweets you want on your home page to be instantly viewable when you turn on your phone or computer? First, go to Preferences, then click on Homepage and choose to Turn All Updates On so that all the items from Twitter and Facebook end up in one place for quick viewing.

- If you haven't changed your ringtone yet, there are a couple of options through which you can do this easily. First, choose Mail, Contacts, and Ringtone on the iPhone home screen and choose an mp3 ringtone from your computer or take one right off iTunes. Then select your favorite song as your new ringtone.

Chapter 16: Maintenance And Battery Replacement Options

How To Fix Battery Life Issues In iPhone 14

Launch Your iPhone

Most of the time, an app or service runs in the background while completing some of the necessary duties. While these programs or services are required to do some crucial activities, they may significantly reduce battery life. Restarting the iPhone is the most popular approach to stop these background apps from running.

So, turn off your iPhone while holding the power button, wait two to three minutes, and then turn it back on. The problem should be resolved.

Turn Off 5G

The iPhone 14 phones may still have battery life drains even though they have stronger processors and can handle 5G better. So, the best action is to turn off 5G and see whether the battery life has improved.

To do that:

- Open Settings app.
- Now head over to "Cellular" > "Cellular Data Options" > "Voice & Data."
- Tap on "LTE," and you are good to go.

Use Low Power Mode

Like Power Saver on Android phones, Low Power Mode is a battery-saving feature that comes standard on iPhones. This function ensures that apps are managed when your iPhone's battery is low so that very little power is consumed, and you can continue using the device for at least a few hours before connecting to a charger.

It typically disables some of the background apps and services to ensure nothing uses the battery beyond what is desired. Here's how to enable this function:

- Launch the Settings app.
- Now navigate to "Battery."
- The "Low Power Mode" toggle switch is located here.

iPhone Reset

Resetting your iPhone is your final option if nothing else has worked for you. Any uncleared cache or useless apps that may be using more battery than others can be removed by doing a reset. Therefore, when you reset your iPhone, these items will be erased, and the problem should be resolved.

To reset your iPhone,

- Launch the Settings app.
- Navigate to "General" > iPhone "Transfer or Reset" > "Reset All Settings" > "Reset."

➤ Once you've entered your password, the iPhone will restart itself.

Battery Replacement Options

If your iPhone 14 battery is starting to lose capacity, there are a few replacement options. Apple offers an iPhone 14 battery replacement under warranty only when the battery is down to 80 percent of its original capacity. Usually, an iPhone battery will function optimally from 18 to 24 months, but if it's getting too worn out, it'll take longer to charge and run slower. If this happens to you, read on for tips on how to find a suitable battery replacement for your model.

First, consider the cost. A replacement battery for your iPhone 14 costs $99 and is covered under Apple's warranty. Apple has historically fixed battery issues under warranty, but it is essential to keep in mind that battery capacity starts to diminish once the battery reaches 80%. This means that the battery has gone through many charging cycles and lost its capacity.

If you can't afford a replacement battery, consider AppleCare+. AppleCare+ has various plans to help you with battery emergencies, including battery replacement. This program covers replacements for up to three years. In addition, you can choose to pay monthly or annually for AppleCare+ or opt for one of the many other protection plans available.

Another option is to take your iPhone to an Apple store. This service is fast and convenient and only requires a small fee. However, it's important to note that battery replacement services are not available at every Apple store. If you live in an area with an Apple Store, you can schedule an appointment at the store to get the battery replaced. Otherwise, you can mail your phone to an authorized Apple repair center.

You may also consider buying a new phone.

Chapter 17: FAQs

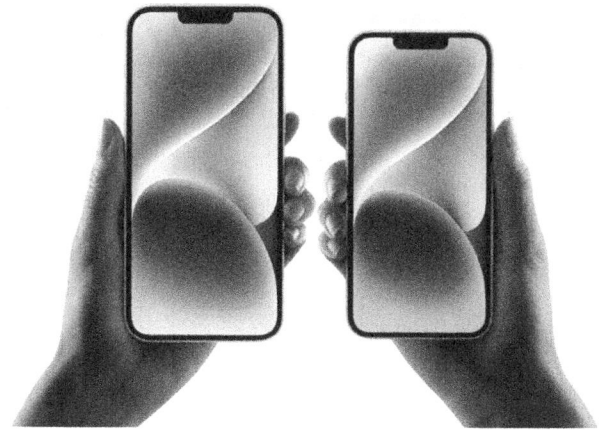

What is an iPhone?

Apple released its first iPhone in 2007, and since then, it has become one of the most recognizable brands in the world today. The phone comes loaded with various applications useful for anyone from seniors to kids. It also provides easy access to email and other internet-based services.

Is there tech support for seniors?

Many seniors feel like technology is very hard to use and will make them look stupid or out of touch in front of others. The best way to get tech support is to have someone else do it or have a friend or family member do it. But people still need to know where to go for help. The internet provides you with how-to guides on YouTube and online articles that can give basic information on how to use different apps and functions. For example, a spell checker is always available if the person isn't sure about the spelling of the contents of an email address. Siri can also be very helpful, as most people who use the iPhone are familiar with her assistance.

Is it easy for seniors to use?

The senior doesn't have to be a tech wizard or have tons of tech-savvy to own an iPhone. There are different iPhones available for purchase to suit different needs.

What apps will help the seniors?

There will never be a right or wrong answer to this question because everyone's needs are different. It's essential to find out what their needs are so you can find apps that will help them more specifically. For example, if they are used to listening to music, they may want to get a phone that will let them listen to music through their headphones.

How Can They Make iPhones More Accessible?

You can't expect a senior to become familiar with all of the parts and features of an iPhone if they have just received it. So, you need to give them time and resources to understand and use the iPhone. You should provide the senior with everything, including charging cords, headphones, extra batteries, cords for cars or other devices, computers, etc... If you don't have those things, you can always go online and get them from places like Amazon. That way, the senior doesn't have to head to a store and bring all those things.

How to find a lost iPhone?

It's important always to know where your senior is located so you can be contacted if they get lost or lose their phone. They shouldn't worry either because it's easy for someone to find a phone if they need help. First, you need to enter the last digits of the person's phone number into something called an Amber Alert app or website. Once that happens, you will have one less thing to worry about, and your senior will be able to be found much more easily.

Does iPhone have more gimmicks than useful technology?

We can answer this question in two ways. For example, an iPhone can have many more gimmicks than useful technology if they aren't familiar with the iPhone and its features. But, when a person has been using an iPhone for years, it is often filled with a lot of creative technology that helps them every day. Those people should also know their phone's features to avoid missing out on anything. Everyone's needs are different, so everyone's experiences will also be different when using an iPhone.

What are some features that iPhones don't have?

Many seniors aren't as familiar with technology and iPhones, so they may not know everything about their phones. However, many of the features that are on an iPhone are very useful for seniors or for anyone for that matter. They can help seniors communicate with others, stay active, keep themselves in good health, and much more. The most popular apps include Facebook, iTunes, Gmail, and more. In addition, many apps can be used to keep seniors in touch with their friends and family.

Is iPhone useless if seniors don't use email?
It shouldn't be too difficult to convince your grandmother to purchase an iPhone if she doesn't have an email. In many cases, seniors prefer a simple phone that can access text messages over the hassle of downloading and installing an app on their computers. Also, if you use an iPhone as a second phone, the seniors can text you there instead of calling. This will probably help save money since they don't need to pay long-distance fees or drive to a library to check email.

Should you buy a phone case?
The answer here is yes and no. It is an excellent idea for seniors to buy a phone case because keeping their phones correctly protected from damage can be difficult. It is also a good idea because many seniors might find that the phone slips out of their hands or falls on the floor more often than they would like. However, we would not recommend buying a custom case made specifically for your iPhone. This can be very expensive, depending on your chosen material and design. Instead, we would opt for cases made of materials with enough grip and strength to protect your phone in case of an accident.

Does an iPhone work with hearing aids?
Many people may need help from family or friends when using their iPhone with a hearing aid. They should try moving their phone closer to the hearing aid and ensure that the volume is turned up on both devices. If they aren't able to hear, then they should try turning off any background noise, such as music on their phone or in public places like restaurants, banks, salons, etc. They can also try having someone read their texts to them if they aren't able to read it on their own.

Can older adults use iPhones?

Yes, older adults can use iPhones, and it would be a great idea to start using them. Many seniors are starting to integrate tech into their lives because it's something they need to stay in touch with others. If you were only to use an iPhone with a senior, you would need to ensure that they can use the phone without assistance from other people.

Hearing loss among adults is rising, and hearing aids are becoming more popular. In the future, hearing-aid technology could include a built-in hearing system that works with your phone and other equipment.

Will the iPhone 14 fold?

Likely not. Honestly, we can't say for certain whether a foldable iPhone will at any point come around, but it is possible in the following couple of years.

Will iPhone 14 be waterproof?

Indeed, the iPhone 14, iPhone 14 Plus, iPhone 14 Pro, and iPhone 14 Pro Max are waterproof. They're appraised IP68 under IEC standard 60529. This implies that they can be lowered to the greatest profundity of six meters (19.7 feet) for up to thirty minutes.

What distinguishes the iPhone 14 from the iPhone 13 most significantly?

Even though it could seem that the iPhone 14 and iPhone 13 are identical, Apple has made a few modifications to set the two apart from one another. The processor, battery, and camera are where the largest disparities exist. Compared to the iPhone 13, the iPhone 14 has a better camera, battery life, and an enhanced processor. The Apple Crash Detection function, which alerts authorities if the user is in an accident and is absent from the iPhone 13 model, is also included in the iPhone 14.

What iPhones support Dynamic Island?

Dynamic Island is a feature of the iPhone 14 Pro and iPhone 14 Pro Max. Due to their wider notches, the iPhone 14 and iPhone 14 Plus don't have Dynamic Island. Dynamic Island is not available on the iPhone 13 or earlier iPhones, either. In addition, although Dynamic Island is a software feature, not all iPhones running iOS 16 will have access to it.

Which phone models support satellite connectivity? What is the process for the iPhone satellite feature?

The iPhone 14 series included the capability for satellite communication. The iPhone antennae can now link to satellites thanks to the software created by Apple. You may access an Emergency SOS using Satellite UI on iPhone 14 devices. You may use this function to send messages and manually share Find My Location with your contacts when you lose cellular or Wi-Fi access. The iPhone will lead you in the direction you must hold the phone, but for this to operate, there must be a clear sky. Even though Apple has reduced the message sizes, depending on how clear the sky is, it would still take between 1-15 minutes to send a message. Your medical ID will also be transmitted along with the mail. You also cannot send anything else at all. Instead, you must answer a few limited multiple-choice questions.

The message will initially attempt to be delivered to the carrier provider. Still, if that attempt is unsuccessful, it will be forwarded to an Apple relay center (with live agents) that will then send it to the telecom operator. Emergency SOS by Satellite UI supports three languages: American English, American Spanish, and French Canadian. The only supported character set is Latin. It is anticipated that it will be made available as part of a subsequent iOS update, maybe iOS 16.2, in certain countries, including the US.

Which iPhones are crash detection compatible?

All versions of the iPhone 14 support the new Crash Detection function. They have a new dual-core accelerometer with capabilities of detecting up to 256Gs of G-force. They also have a new high dynamic range gyroscope. These allow the iPhone to detect whether you are in a vehicle accident (from the front, back, side, or rollover). You are first prompted to connect to an emergency service. Then, it immediately notifies the emergency contacts and rescue agencies of your precise position if you don't react within 10 seconds.

What differentiates the iPhone 14 Pro from the iPhone 14?

The iPhone 14 Pro boasts Dynamic Island, a brighter 120Hz display that supports Dolby Vision, an Always-On Display (AOD), greater battery life, a 48 MP primary camera, and the A16 Bionic SoC compared to the iPhone 14.

Conclusion

Having understood the iPhone 14 and its key features, it is hard to say if it is not the best cell phone on the current market. Some people criticize the iPhone as a device that does too little and costs too much for what you get. But Apple fanatics will tell you that there are less costly smartphones, and that's true, but it's also true there are others that cost more.

However, portable computers or mobile phones like iPhone 14 have come a long way since 2007. Back then, it was expensive to access an internet connection, let alone for most people to own an iPhone. Although the internet is now ubiquitous and has been incorporated into all new smartphones, the Apple iPhone remained one of the few that offered access to many applications and programs.

In addition to their proprietary apps (such as iBooks, FaceTime video chat on calls, and iTunes), iPhones can connect directly to other devices like iPads, iPods, and even Mac computers. So, if you can imagine a particular application, chances are that it already exists, and you have everything you need to make it happen on your iPhone 14.

The rest is history: the iPhone has changed the way people interact with each other in real life. The best thing about this is that you can use your iPhone 14 on the go or in normal conditions and still get a top-notch experience.

Apple's latest iPhone lineup is imposing. The company leans into its silicon lead to making significant improvements in cameras, and it uses its design team to develop new ways for people to interact with their phones. Its always-on display is a vast improvement over its competitors, and it performs well in low-light conditions. Lastly, the camera improves on its predecessors with a bigger sensor and larger pixels.

The iPhone 14 is an excellent choice for seniors. It has improved cameras and has satellite SOS for emergency assistance. The battery life of the new iPhone is also better. Despite inflation, consumers still prefer high-end products. This model offers a bigger screen, an upgraded camera, and an improved battery. The iPhone 14 is your way to go.

Made in the USA
Columbia, SC
29 June 2023